FOSS Science Resources

Animals Two by Two

Developed at
The Lawrence Hall of Science,
University of California, Berkeley
Published and distributed by
Delta Education,
a member of the School Specialty Family

© 2016 by The Regents of the University of California. All rights reserved. No part of this book may be reproduced or transmitted in any form or by any means, electronic or mechanical, including photocopying or recording, or by any information storage and retrieval system, without permission in writing from the publisher.

1487694
978-1-62571-420-6
Printing 1 — 6/2015
Standard Printing Company, Canton, OH

Table of Contents

Fish Same and Different.................3
Fish Live in Many Places 10
Birds Outdoors 20
Water and Land Snails 29
Worms in Soil........................ 37
Isopods 48
Animals All around Us............... 55
Living and Nonliving 67
Glossary 87

Fish Same and Different

Goldfish and guppies are **fish**.
Where do you find them?

3

They live in ponds and lakes.
Other fish live in the ocean.
Where else do fish live?

Fish are **animals**.

All animals have **basic needs**.

Fish need food and water.

They need space and **shelter**.

They need **oxygen**, too.

Fish have eyes and a mouth on their heads.
They have **scales**, **fins**, and **gills**.
Do you see them?

All fish **breathe** through gills.
Fish breathe oxygen from the water.

7

Color, shape, and size are **characteristics** of fish.

How are these fish the same?
How are they different?

Fish Live in Many Places

Fish need to live in water.
Some fish live in fresh water.

Rivers and streams are fresh water.
Ponds and lakes are fresh water.
Do you see fish here?

11

Other fish live in salt water.
The ocean is salt water.
Many different kinds of fish
live in the ocean.

Some marshes are salt water.
Fish that live in salt marshes
need salt water to live.

Aquariums are places to see fish.
At an aquarium, fish live in large, glass, indoor tanks.
These fish live in a saltwater tank.

Pet stores are places to see fish in small tanks.
Have you seen these fish?

Your class has a fish tank.
Is it fresh water or salt water?
Goldfish live in fresh water.

Fish live in many places.
Some fish live in natural places.
Other fish live in human-made places.

Some fish eat plants.
Some fish eat animals.
Some fish eat both plants and animals.

Fish get what they need from where they live.

What are the basic needs of fish?

Birds Outdoors

Where do you find birds?
Birds fly in the sky.
They sit on tree branches.

They sit on power lines and fences.
Can you find a bird outdoors?

Birds are animals.
Where do birds find food to eat and water to drink?
Birds need air and shelter, too.

Birds have feathers and wings.
What else do you see?

Birds have a beak to get food.
Some birds eat fish.
Other birds eat **worms** or insects.

24

Some birds eat plants and seeds.
What else do birds eat?

Birds build nests for shelter.
Female birds lay **eggs** in a nest.
What will hatch from the eggs?

Look at the shapes and colors of these birds.
Look at their feathers, eyes, and beaks.

How are these birds the same?
How are they different?

28

Water and Land Snails

Snails are animals.
Where do you find snails?
Some snails live in water.
They live in fresh water
or salt water.

Some snails live on land in **moist** places.

A garden is one place
for land snails.

31

Snails have a shell and a **foot**.
Can you see the shell?
Can you see the foot?
The foot helps the snail move.

All snails have a mouth.
Snails scrape food into
their mouth.

Snails also have **tentacles**.
Look on the snail's head.
Do you see the tentacles?

The tentacles **sense** food, air, and light.
What else do tentacles sense?

35

How are these snails the same?
How are they different?

36

Worms in Soil

Worms are animals.
Where do you find worms?
These worms live in the soil.
They get water and food from the soil.

Worms need air, space, and shelter, too.
They get what they need from the soil.

Some worms live in **compost** piles.
Can you see the worms on the plants and soil?
What are they doing?

Worms eat dead leaves and fruit.
They turn old plant parts into soil.
The new soil helps young plants grow.

Some worms **burrow** deep in the soil.
Others stay near the top.

Damp leaves lie on top of the soil.
Worms stay moist under the leaves.
They also find food there.

Worms have **segments**.
Segments help the worms move.

43

Adult earthworms have a **clitellum**.
The clitellum makes a case for worm eggs.
Baby worms grow inside the eggs.

A worm's skin is moist and smooth.
Worms breathe through their moist skin.

Some worms live on land.
Some worms live in fresh
water or salt water.

46

How are these worms the same?
How are they different?

47

Isopods

Isopods are animals. Where do you find isopods? These isopods live on land. Look for them under leaves and rocks where it is moist.

Isopods eat leaves, seeds, and fruit.
They help make soil.

What other animal makes soil?

Isopods have **antennae** and legs.
They have segments, too.
Some isopods move fast.

All isopods breathe with gills.
What other animal has gills?

These isopods live in water.

53

How are these isopods the same?
How are they different?

Animals All around Us

Animals are all around us.

Animals look the same and different.
How are these animals the same?
How are they different?

Lizards and snakes are covered with scales. Can you see the scales?

Do you see another animal with scales?

Fish have scales.
Fish are different from lizards.

Other animals are covered with fur or hair.
Fur and hair keep animals warm.

People have hair to keep them warm, too.

How is the fur different on each of these animals?

61

Frogs and salamanders have smooth, moist skin.
They move on land and water.

Earthworms have smooth, moist skin, too.
Are earthworms the same as frogs?
No, they are different.

Animals look the same and different. They live in many places.

They eat, drink, and care for babies.
All animals have basic needs.
They need food, water, air, space, and shelter.
Do these animals have what they need?

What are the basic needs of animals?

66

Living and Nonliving

Look at the garden.
What is **living** in the garden?

The garden has many plants.
Animals live here, too.

Living things have basic needs.
Living things grow and change.

Do you see the young plants?
They are the **offspring** of full-grown plants with flowers.
The young plants will grow up to look like their parents.

What is living in this garden?
What is not living?

Rocks are not alive.
Water is not alive.

Rocks and water are **nonliving** parts of a garden.
Rocks and water do not grow.
They do not have offspring.

Look what lives at the edge of the ocean!
What living things do you see?
What nonliving things do you see?

73

Many animals lay eggs.
The eggs hatch into babies.
Then, the babies eat and grow.
Soon, they might have offspring
of their own.

Beetles and butterflies lay eggs.

Fish and birds lay eggs.

Mammals do not lay eggs. Their babies grow inside the mother.
When they are born, baby mammals need a lot of care.

Mother mammals feed their babies milk. Where do piglets get milk?

Sunflower Life Cycle

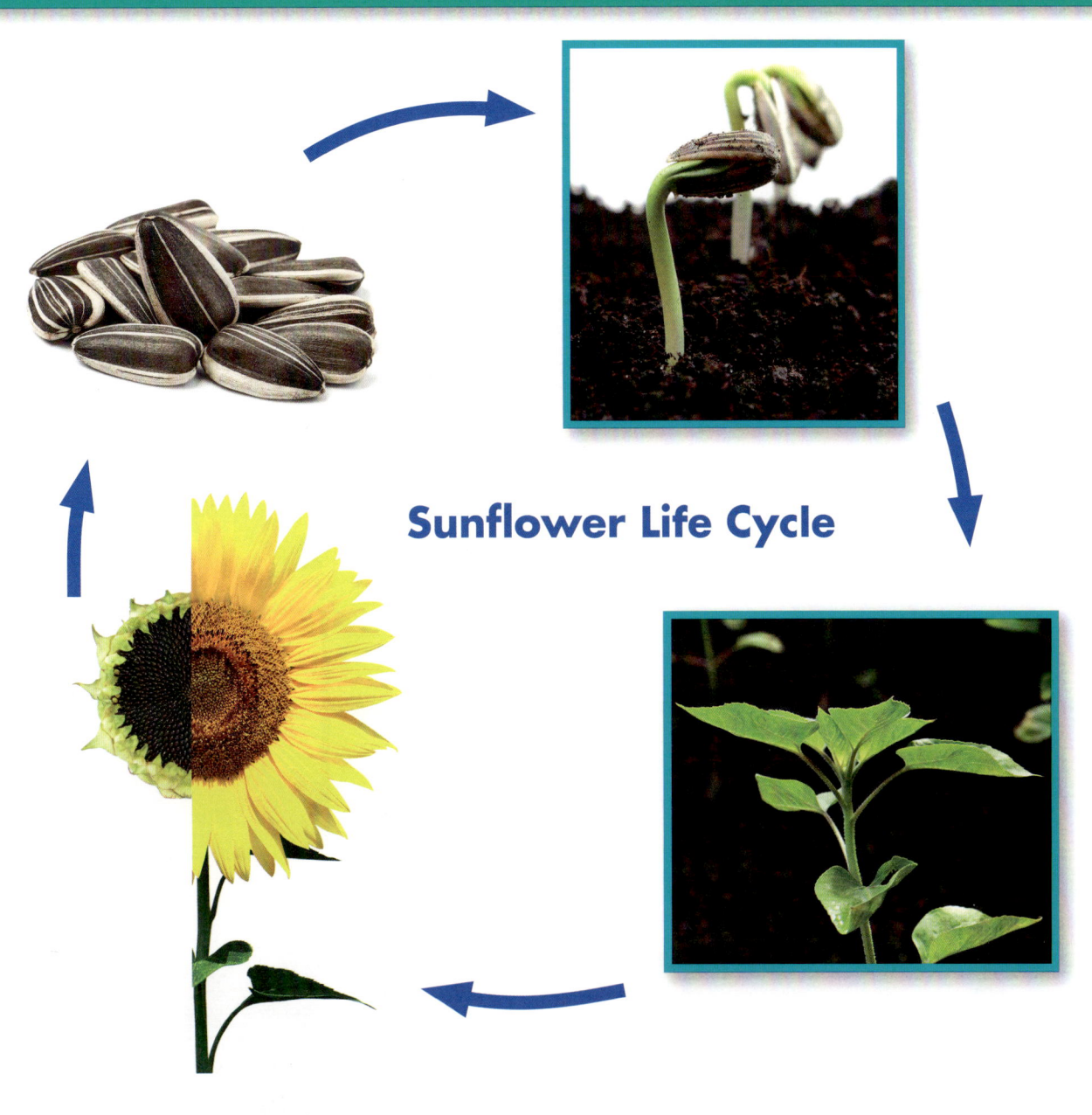

Plants don't have babies.
Some plants produce seeds.
The seeds form in flowers.
Each seed holds a tiny living plant.

These seeds need soil and water.
Then, the little seeds can grow into big pumpkin plants.
What will be inside the pumpkins?

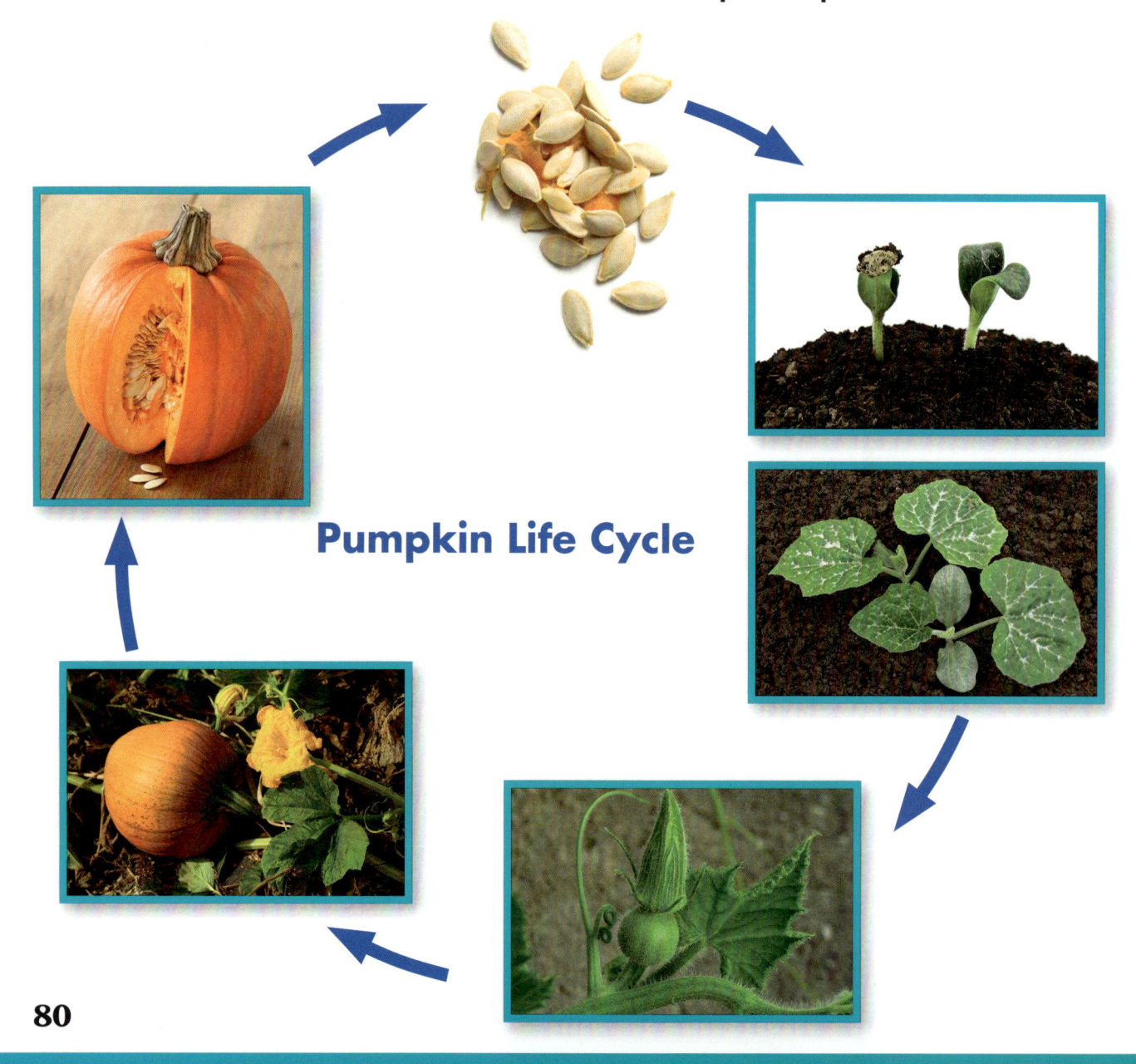

Pumpkin Life Cycle

Look at the photos on this page.
What is living and nonliving?
Why do you think so?

81

What is living and nonliving?
Why do you think so?

82

What is living and nonliving?
Why do you think so?

83

What is living and nonliving?
Why do you think so?

What is living and nonliving?
Why do you think so?

85

Look at the pictures.

Name two things that are living.

Why do you think so?

Name two things that are nonliving.

Why do you think so?

Glossary

animal a living thing that needs air (oxygen), water, food, space, and shelter **(5)**

antenna (plural **antennae**) a feeler on the head of an isopod or insect **(51)**

basic need something that plants and animals need to survive. Plants need air (oxygen), water, nutrients, space, and light. Animals need air, water, food, space, and shelter. **(5)**

breathe to take in air (oxygen) **(7)**

burrow to make a hole for shelter **(41)**

characteristic how a plant or animal looks or behaves **(8)**

clitellum a band around an earthworm's body **(44)**

compost dead plants and leaves **(39)**

egg what some animals hatch from **(26)**

fin what a fish uses for swimming **(6)**

fish an animal that lives in water **(3)**

foot what some animals and people use to move around and stand on **(32)**

gill what some animals use for breathing **(6)**

isopod an animal with a segmented body and seven pairs of legs for movement **(48)**

living alive. Able to grow, change, and produce offspring **(67)**

mammal an animal that has live offspring and produces milk **(77)**

moist a little wet or damp **(30)**

nonliving not alive **(72)**

offspring a new plant or animal produced by a parent **(69)**

oxygen a gas found in air and water **(5)**

scale what covers the body of a fish and other animals **(6)**

segment a section of a worm's or isopod's body **(43)**

sense to become aware of something by seeing, smelling, hearing, touching, or tasting it **(35)**

shelter a safe place where animals live. A shelter protects an animal from weather or other animals. **(5)**

snail an animal with a shell, tentacles, and a foot **(29)**

tentacle a feeler on the head of a snail **(34)**

worm a soft, moist animal with a segmented body **(24)**